Anonymous

A Child's first Book about Birds

.

Anonymous

A Child's first Book about Birds

ISBN/EAN: 9783337331009

Printed in Europe, USA, Canada, Australia, Japan

Cover: Foto ©berggeist007 / pixelio.de

More available books at **www.hansebooks.com**

A CHILD'S

FIRST BOOK ABOUT BIRDS.

BY

A COUNTRY CLERGYMAN.

> I shall not ask Jean Jaques Rousseau
> If birds confabulate or no ;
> 'Tis clear they were always able
> To hold discourse at least in fable.
>
> COWPER.

LONDON: JAMES HOGG & SONS.

CAMDEN PRESS, LONDON.

To

CARREY AND BLANCHEY,

THIS LITTLE BOOK

IS INSCRIBED

BY THEIR FATHER,

THE AUTHOR.

CONTENTS.

THE TURKEY AND THE GAME-COCK.

THE TURKEY AND THE GAME-COCK.

In a farm-yard in Norfolk, a county famous for turkeys, there was one day a dark grey cock-turkey, it is a common colour among these birds, he had a tuft of black hairs on his breast, the top of his head was bare and red, and of course there was a bit of bluish skin hanging down from his bill, for no one ever saw a cock-turkey without such a bit of skin. Well, this turkey, who considered himself the king of the poultry-yard, was strutting about with his tail spread out, and it looked like a fan when it is open;

and as he walked the ends of his wings touched the ground. Now, by chance, a game-cock, whose plumage was dark brown and gold, being in search of food, came to that part of the yard. The turkey's neck in a moment turned from pale blue to red, a sure sign he was angry. The cock was not so big as the turkey by a good deal. His head was small, and so indeed is the turkey's for his size; he had sparkling eyes, and his comb and wattles were deep red, which might be the cause of the turkey being angry, for anything of that colour puts a cock-turkey into a passion. The feathers of the cock's neck were long, and those in his tail drooped in the form of an arch. His step was firm and he had a bold look. It is well known that no bird has more courage.

The turkey having first made that low, hollow sound which cock-turkeys make when any one comes near them, thus addressed chanticleer.

T.—Fellow, what business have you here?

C.—I have as much right to be here, Mr. Strut, as you have.

T.—Pert coxcomb, go back to your dunghill; the proper place for birds of your feather.

C.—I shall stay here as long as I please, in spite of your red rag, and your angry looks.

T.—Do you presume to talk in this manner to one who has two tails.

It is true that cock-turkies have two tails, but only one of them opens like a fan; the other, which is underneath this, is short and small.

C.—I care not for any bashaw with two tails; but although I have only one tail, I have not a bald head.

T.—You talk of a tail! you, who, when trimmed for fighting, look more like a scare-crow than a bird.

C.—That is no disgrace to me, though it is to them who, for a bad purpose, deprive me of what nature intended to be an ornament. But you turkeys are such cowards I could frighten a whole flock of you by shaking a few peas in a bladder.

When turkeys are driven in flocks, this method is used to get them along the road. They mistake the noise of the peas for thunder, of which they are very much afraid.

T.—Every cock can crow on his own dunghill.

C.—You are such a silly creature that

any old woman in a red cloak puts you in a rage.

T.—And you are a fool, for I have seen you pick up pebbles, and, thinking they were grains of corn, swallow them.

The turkey ought to have known that other birds do the same, and they have a reason for it; the small stones which they swallow from time to time help to digest their food.

C.—You are a monster, for if not prevented, you break the eggs on which your mate is sitting.

T.—You are a common brawler, and beneath my notice.

C.—Cock-a-doodle-doo!

As the cock said this he clapped his wings.

T.—Begone this instant, saucy crower; go and scratch the dunghill for your dinner; if you stay here you shall feel the strength of my wings.

C.—Come on, Sir Brag, and you shall find that my spurs are both longer and sharper than those on your ugly black legs. You forget that I am a true English bird, but you turkeys came to this country from foreign parts.

The cock had doubtless heard that turkeys were brought to England from America.

Both the birds were now in a great passion. The turkey's neck was as red as crimson, and the wrath of the cock was seen in the fierceness of his eye. Their dispute having by this time reached the ears of the other poultry in the farm-yard, turkeys, geese, ducks, hens, and pigeons, flew to see what was the matter. Some of them took the turkey's part, and others sided with his adversary. The noise now became much louder than before. The turkeys gobbled, the cocks crowed, the geese and hens cackled, and the ducks quacked. At last the uproar brought out the farmer's wife, who quickly sent them all about their business

THE SWAN AND THE GOOSE.

The Swan and the Goose.　　　　　P. 19.

THE SWAN AND THE GOOSE.

UNDER the arch of a rustic bridge which spanned a clear stream running through a gentleman's park, there was to be seen, one bright summer's day, a swan and a goose having a quiet chat. Now, for fear you should not know a swan from a goose, I will describe each of them, before I tell you what they were talking about. The swan was in full plumage, and therefore could not be less than two years old. Its feathers were snowy white, and its body was also covered with a thick down of the same

colour. Of this down ladies' tippets are
often made. The swan's bill, which was
red, with a little black at the edges and
tip, was about three inches long, and the
legs were quite black. The swan may
be known from all other English birds
by the great length of its beautiful neck,
which it bends in the most graceful
manner. It is the largest of our birds,
and weighs about twenty-five pounds.
The tame swan is larger than the wild
swan, and this is the case with most other
birds, and it is owing to the tame birds
getting more food than the wild ones.

The goose was standing on one leg, a
favourite posture with that bird. This
goose was of a grey colour, with shades
of ash, blue, and brown, in its wings.
White geese, however, are as common as
grey ones. The bill of the goose was a

yellowish red, thick at the bottom, and tapering towards the tip, and its legs were pale red. A full-grown goose weighs from nine to fifteen pounds. It is, however, a bird so well known that there is no need to describe it more exactly.

Now, although the swan was so much larger than the goose, yet it did not molest it in any way. It is much to the credit of the members of the swan family that, notwithstanding their great size and strength, they never ill-use any of the other fowl which frequent the same water with themselves. The gentlemen swans are fierce only when their ladies are hatching; at which times they will not suffer any bird or animal, or even a man to approach their nests without attacking the intruder; it is their way of showing their affection for their wives and little ones.

Let me now relate to you the conversation that passed between the swan and the goose on this occasion.

S.—I am glad, Mother Goose, to have your company under this shady arch this hot day.

G.—I was just thinking, Madam Swan, what a noble sight it must be to see a flock of swans swimming on a lake. 1 have heard it is sometimes seen in Canada.

S.—But a more singular, though, perhaps, not so beautiful a sight, is occasionally seen in Australia—a flock of black swans.

G.—I have been told that a swan can fly at the rate of a hundred miles an hour. Have I been truly informed?

S.—Yes; it is owing to the great strength of our pinions that we are able to fly so fast.

G.—I am not surprised, however, that you prefer swimming to flying or walking, for you move through the water so gracefully and so swiftly. I have seen you swim as fast as a man can walk.

S.—I know ill-natured people say that we prefer swimming to walking because we are ashamed of our black legs. The truth is, we swim with ease, but we soon get tired when we walk.

G.—Allow me to ask, are you ever deprived by your owner of the large feathers in your wings?

S.—Never while we are alive; but I am informed, on good authority, that after we are dead our wing feathers are plucked out, and their quills are made into pens, which sell for as much as three pounds per hundred. I am aware, Mrs. Goose, that your quills are used for the same purpose, but not being so large as ours they do not fetch so high a price. I believe your feathers are sometimes plucked from you while you are alive.

G.—In some parts of England, particularly in Lincolnshire, we are plucked as often as five times in a year. About Lady-day we are first deprived of our feathers and quills, and between that time and Michaelmas we are plucked four times for our feathers only. Even our little ones of six weeks old are not spared; they, however, have only their tails plucked.

S.—We are very fond of fish, but we are often obliged to eat other kinds of food. Pray, what do you chiefly live on?

G.—On corn and herbs ; but if any little fishes come in our way we do not scruple to swallow them.

S.—Although we are supposed to exceed you in beauty, you must be more valuable to your owners than we are, as you furnish them not only with feathers, down, and quills, but also afford them a very nice dinner on Michaelmas-day.

G.—We tremble whenever we hear that day mentioned. When I was a gosling, I used to think that the man who had the care of the flock to which I belonged (he was called the Goozard), was very kind indeed to the full-grown geese, for he actually crammed the food down their throats ; but I afterwards found out it was not altogether from kindness that he forced them to eat so much, but it was to make them as fat as possible, and so to get a better price for them at market.

S.—Our fate is different from yours; swans are never obliged to eat against their will; but cygnets, that is the name given to young swans, are sometimes fatted for the table, and sell at a high price. I lost some of my own brothers and sisters in that way.

G.—I have heard that, like ourselves, swans are a long-lived race.

S.—It is said some swans have attained the age of three hundred years; but such very long lives do not often occur among us

G.—The usual length of a goose's life, if nothing untoward happens to it, is about seventy years, which I am told is the term of life allotted to man; who in that respect, therefore, is no better off than a goose.

S.—Instead of geese being more foolish (as they are generally accounted), I think they are more sagacious than other birds. Your vigilance once saved Rome, then the capital of the world; your affection to your own species, and your attachment to other birds, to animals, and even to the lords of the creation, has been often shown. And besides this, you possess great courage, which is well known to the Russians, who train you as the English do game-cocks, for fighting.

G.—Talking of fighting reminds me of the great strength and courage of your own tribe. A swan has been known to break a man's leg with a stroke of its wing; and I have heard of another that pulled a fox under the water and held him there till he was drowned. And the gentlemen of your family when they do fight about a lady, generally fight until one of the combatants is killed.

S.—Like other birds, we defend our mates and our young ones; but between ourselves there is a good reason why we should avoid a fray, a slight blow on the head will lay a swan dead in an instant.

G.—You are less frequently seen walking on the land than swimming in the water, but it is just the reverse with us.

S.—We can live entirely on the land; but if we are kept from the water for a long time we become awkward, dull, and dirty.

G.—If we were not more on our legs than you are, we should not be able to take such long journeys as we are sometimes forced to take. Perhaps you have

never heard that large flocks of geese
(there was once a flock of nine thousand)
are driven every year all the way from
Lincolnshire to London.

S.—You surprise me. How far do
they travel in a day ?

G.—From eight to ten miles.

S.—I should think many of them must
fall lame on the road.

G.—Those that fall lame are put into
a cart, which serves as a kind of moving
hospital, and follows the flock all the
way.

S.—It must be a tiresome task to make
them walk such a distance day after day,
and to keep them from straying out of
the right road.

G.—To prevent them from loitering by
the way the drivers wave a terrible red
flag over their heads, and then they dare

not stand still; and if they attempt to
stray, the drivers catch hold of their legs
with a hook at the end of a long stick.

S.—The cackle of a goose to my ear is
a very cheerful sound.

G.—I have heard that the Icelanders
rejoice to hear the melodious voices of
your tribe, and it is no wonder, for it is
to them the first signal of the approach of
spring, after their long and dreary winter.

S.—Pray, Mrs. Goose, how many eggs
do you lay?

G.—My number varies from seven to
twelve; they are white; and my nest,
made of hay or straw, is to be found
under the shelter of some building, and
not in the open air.

S.—My nest is made of dry herbs or
reeds, and I place it on the bank of a
lake or river. I lay six or seven white

eggs ; and, like you, I lay one egg every other day. Perhaps you never heard that formerly it was felony to steal swans' eggs.

G.—Before we part allow me to ask, were you always as beautifully clothed as you now are?

S.—Oh! no. When I was a cygnet my only dress was a yellowish down, like that on your own goslings ; but at two years old I put on my present handsome attire

The swan now hearing her mate's call, made a graceful bend of the neck towards the goose and sailed away.

THE GOLDFINCH,
THE BULLFINCH, AND
THE CHAFFINCH.

2

THE GOLDFINCH,
THE BULLFINCH, AND
THE CHAFFINCH.

On each side of an open window, facing the south, hung a bird-cage. In one cage there was a goldfinch, and in the other a bullfinch. The goldfinch's forehead was scarlet; a proof that he was at least two years old, for goldfinches do not have scarlet foreheads before their second year. The top of his head was like a piece of black velvet, his back reddish-brown, his breast white, and the middle of his wings was of a golden hue; it is this that gives the bird its

name ; the tail feathers were black. He
was a native of Kent; the finest goldfinches are found in that county. The
plumage of his friend in the other cage
was almost as gay as his own. The
top of the bullfinch's head was glossy
black, the upper part of his neck and
breast light red, and his dark wings and
tail were spotted with white.

These two little birds were not only
near neighbours, but, which is not always the case with neighbours, they
were also great friends, and spent a good
deal of time in chatting about their
acquaintances, and the things they could
get a peep at from ·their perches ; for,
fortunately for them, the window at
which their cages were hung was put
open in fine weather, to admit the fre
air. This enabled them to look into the

fields, and to receive the visits of such of their relations and friends as lived in a state of freedom.

One bright summer's day, the gold-finch and bullfinch happening to have no visitors, and being tired of singing, they fell into the following conversation :

B.—I observe you are often very busy; you seem to delight in moving the things in your cage from one place to another

G.—We goldfinches are an active race, we cannot be idle many minutes together. The water I drink I draw up from the fountain which you see hanging beneath my cage.

B.—As you are so industrious, I do not wonder that your nests are neater and more strongly built than those of any other English bird.

G.—We do take great pains in building them.   Will you permit me to make one remark respecting your voice?

B.—Pray do.

G.—It strikes me that the notes of your brothers and sisters, who sometimes pay us a visit at our window, are not so melodious as the tones of your own voice.

B.—Your observation is just; and the reason is, because they have never been taught to pipe.   Before I was an inhabitant of this pretty little house—I mean, when I was in a wild state—I had, like them, only three notes, and they were rather harsh.

G.—How were you taught to pipe?

B.—Ah! my friend, your song is the gift of nature; but I should never have sung the songs you hear me sing had I not been taught by art.

G.—I cannot imagine how it can be managed ?

B.—Well, it was in this way: When I was only two months old, I was put into a dark room, where I could see nothing, but I heard the music I now sing. I listened, and listened, and after awhile I thought I would try to sing like it. All at once, the music ceased, the room became light, and then my music master, as I used to call him, gave me some canary seed, which you know is the favourite food of bullfinches.

G.—Did you learn your beautiful song in one lesson ?

B.—Oh ! no; I went on in this way day after day, for several weeks, till I was perfect in the tune. And you may be sure I did my best, for my master never gave me my breakfast till I had sung my song to please him.

G.—What sort of a musical instrument did your music master use?

B.—He called it a bird-organ, but it was only a small square deal box.

G.—Thank you, my friend; I know now how bullfinches are taught to pipe.

B.—Is there any other question you would like to ask me?

G.—Pray, are you related to the wag-tail family? I take the liberty to ask this question, because I observe that you frequently move your tail up and down.

B.—Our family is not related to the wag-tails. It is merely a habit of ours, which, perhaps, we learned from them. Speaking of habits, I observe you always roost at night on the topmost perch in your cage.

G.—Yes, that is one of our habits. What a very fine day it is. I suppose, if you were at liberty, you would spend it in the woods, the resort of bullfinches, I am told, during the summer.

B.—I dare say I should follow the example of the other members of my family; but I am sure I should come back to my cage as soon as the cold weather set in.

G.—Confinement, at all events, does not shorten the days of goldfinches, for one of our family lived nearly twenty years in a cage; and yet, in our natural state, we are social, and live in flocks.

B.—There is this advantage in living always indoors : we are able to sing two or three months longer than we could sing if we were living in the woods.

G.—For my part, I am willing to remain the tenant of this cage as long as I have you for my friend and neighbour.

The friendship of these birds for each other seemed to be too strong to be broken; but alas! among those who occasionally paid them a morning visit was one Mr. Chaffinch, a tattling busybody, who, having no business of his own to attend to, made it his daily occupation to meddle with the affairs of other people. He was smartly dressed; he wore a suit of a chesnut colour, mixed with red and olive, and he had a white spot on each shoulder. Just at this moment this Paul Pry dropped in; and, seeing the goldfinch viewing himself in a

little bit of looking-glass, fixed at the
end of his cage, though all the time he
thought he was looking at another goldfinch, their tattling visitor said in a whis-
per to the bullfinch, "A word in your
ear;" and put his bill close to the bullfinch's ear. What he said was heard by
no one but the bullfinch. Mr. Chaffinch
then talked aloud about the weather and
the crops, till he saw the bullfinch's back
was turned, when he whispered something in the goldfinch's ear; and having
done this he took his leave. Any one
would at once have seen, by their manner towards each other, that this tattler
had made a breach between these two
old friends. They did not resume their
conversation, or otherwise we should,
perhaps, have learned more of their habits
and instincts.

THE MAGPIE AND THE WOOD-PIGEON.

THE MAGPIE AND THE WOOD-PIGEON.

ONE winter's morn, at sun-rise, a wood-pigeon was winging his way through a copse. When the sun rises wood-pigeons leave their nests. A magpie, on a tree hard by, began at that moment to chatter. Now, the wood-pigeon knew by this that there was about to be a change in the weather; so, instead of flying any farther, he closed his wings, and settled on a bough near the magpie.

W. P.—I thank you for warning me that it is about to rain. I shall return

home, as I do not like to get my feathers
wet.

M.—I shall be very glad of your company here, on the outside of my nest;
pray, perch yourself by my side, and let
us have a little chat, for you may take
my word for it there will be no rain just
yet.

The wood-pigeon flew up to the magpie.

W. P.—Dear me, your nest is covered
at the top.

M.—Yes; it is not open, like the nests
of most birds; the way into it is through
this hole in the side. This side entrance
makes it more secure against the attacks
of our enemies the hawks, and other
birds of prey.

W. P.—I observe that you magpies
build your nests at the top of lofty trees;

but we wood-pigeons build in the hollow trunks of old trees. We differ from you also, I believe, in the number of our eggs, as well as in their colour.

M.—We lay eight eggs of pale green, with black spots.

W. P.—We lay only two, and they are quite white; but then we have two broods in the year, and you have only one.

M.—I hope I shall not offend you by remarking, that now I am close to you, I find that the feathers of those pigeons who live in houses built on purpose for them, are brighter and gayer than yours.

W. P.—We, who live in the woods, although not so finely dressed, are larger than our cousins the tumblers, the pouters, the shakers, and the fantails, to whom you allude.

M.—But your bill and your legs, I see, are pale red, like theirs, and your claws are black, which is the colour of their claws.

W. P.—Yes; all the, members of the pigeon family have the same coloured claws, legs, and bills. Allow me to ask what do you measure from the tip of your bill to the end of your tail? I measure just fourteen inches.

M.—My length is eighteen inches; but although we have the advantage of you, in length of body, you greatly excel us in swiftness of flight. I have heard strange stories of pigeons that are able to fly hundreds of miles without stopping to rest, and that they can carry a letter under their wing all the way.

W. P.—They are called, on that account, carrier-pigeons. We wood-pigeons, however, do not fly far from home, and we never carry a letter for any one.

M.—The postage of letters is now so trifling it is not worth any one's while to turn letter-carrier, unless employed by the Post Office.

W. P.—Since I have been sitting by your side I have been looking at your plumage, and I see that your head and neck are quite black; and it seems to me that if it were not for the patches of white on your wings and heart, and also if your tail, which is shaped like a wedge, were a little shorter, you might be taken for a crow.

M.—By that observation I find you are not aware that the crows are of the pie family.

W. P.—I have heard that you magpies can talk in a language foreign to birds. Pray, is that true?

M.—[Opening his bill very wide.] My tongue, you see, is black; but that is not the cause of my being able to speak a foreign language. Nature has given us magpies the power of imitating many of the sounds which we hear. An unfortunate magpie is now and then caught by idle boys, who teach it to speak; and they are so cruel as to slit its tongue with a thin sixpence, because they think it will then be able to speak more plainly; but that is not the case, for I can speak as well as any magpie ever speaks, and yet my tongue is in its natural state.

W. P.—Be so good as to let me hear you speak two or three of these foreign words.

M.—Poll! Jack! Margo!

These words were uttered in a very harsh tone, and as neither the magpie nor the wood-pigeon knew the meaning of them, they were no wiser than before.

W. P.—Are not you magpies fond of setling on the backs of cows and sheep?

M.—Yes; but it is not for the sake of a ride; the truth is, we are very fond of the insects we find on their backs.

W. P.—You are charged with playing tricks, and being very cunning.

M.—I will show you what my nest contains. Take a peep into it. The bed which you see at the bottom is made of wool and hair, which we pick from the backs of cows and sheep; our young ones lie on this bed. In this corner I keep my money. [There were several shillings, sixpences, penny-pieces, and half-pence.]

Here is my plate. ·[Where the magpie pointed, the wood-pigeon saw some silver spoons.]

And here is my wardrobe. [There were a good many pieces of linen, silk, and lace.] Do you know any other bird whose nest contains such riches?

W. P.—I do not, indeed.

M.—Shall I put you in the way of growing as rich as I am?

W. P.—I must decline your offer.

M.—Do you not wish to have money, plate, and such fine things as I have just shown you?

W.P.—We may be both rich and good, but our riches must be got by honest means, and be used for the benefit of others as well as for our own advantage.

M.—I am afraid you have heard something to my disadvantage.

W. P.—I am sorry to say I have been told, on good authority, that you have grown rich by dishonest practices. I would rather it should be said of me that I was poor and honest, than that I was rich, but also a thief.

This observation showed that the wood-pigeon had been brought up in good principles. The magpie hung down his head and was silent, and the wood-pigeon flew away and got back to his own nest just as it began to rain.

THE OWL AND THE NIGHTINGALE.

The Owl and the Nightingale.   P. 60.

THE OWL AND THE NIGHTINGALE.

ONE calm, soft night in spring, an owl sat on the roof of an old barn. The moon shone so brightly, you might see that the colour of its head, back, and wings was pale chestnut, with gray and brown spots, and its breast was white. Like all other owls, its legs had feathers down to the toes. As there was light enough to show the colour of its plumage, of course its noble round white face, with the strong hooked bill and large round eyes, were also plainly seen. The owl had probably supped, as it seemed to be in a musing mood, rather than on the look out for prey; although night is the feeding time of the owl tribe.

Sounds more liquid than the notes of a flute now broke the silence of the night. The owl, although not one of the long-eared ones, showed that he had an ear for music, for he quickly made his way towards the thicket which held the songster, and so pliant were his wing-feathers, that he made no noise as he flew thither. His piercing eye soon spied out a little bird about six inches long; the upper part of his body was of a rusty brown colour, tinged with olive, the under part pale ash, and it had a whitish throat. The owl at one glance knew it was a nightingale. **The** night warbler sang for a long time without stopping, but at length pausing to take breath, the owl thanked him for his song. They then fell into discourse.

O.—I wish I had your fine voice; I should then lose my nick-name of screech-owl.

N.—You forget that you are regarded as an emblem of wisdom, while I am looked upon merely as a good singer.

O.—I believe you prefer, as we owls do, a quiet life to living in a noisy crowd, which is the delight of my neighbours, the rooks.

N.—We are shy and timid; some people, indeed, say we are very simple creatures; but they do not speak the truth when they declare that we are so proud that we will not sing until all the other songsters of the grove are gone to rest, that we may be heard alone. It is true we sing chiefly at night, but our song is also to be heard in the day time.

O.—I take your word for it; but as I seldom stir abroad till after sunset, I never hear you sing but at night.

N.—I wonder you shun the day-light. We nightingales enjoy both the sun-shine and the moon-light.

O.—The sun is too powerful for my eyes; besides, had you seen how I was treated not long ago, when I was tempted to take an airing in the day time, you would not be surprised that I keep to my old hours.

N.—Did you meet with. any thing to alarm you?

O.—I will tell you what happened; as soon as I made my appearance in the day-light I was hooted at and mobbed by a parcel of paltry sparrows, jays, and finches ;. I was never so insulted in my life.

N.—I am very sorry to hear it.

O.—And there are scandalous stories told of us. The farmers accuse us of sucking pigeons' eggs and killing game, whereas our chief food are rats and mice ; which, but for us, would soon eat the farmers out of house and home. And besides this, some foolish people call us heralds of death, although we have no more to do with dying persons than you have.

N.—I hope you will not take offence at the question I am going to ask.

O.—Pray let me hear it.

N.—Is it true that owls snore while they are asleep?

The owl, in spite of his grave face, could not help laughing.

E

O.—While asleep we certainly make a hissing noise, which some folks perhaps may call snoring.

N.—I shall make a point of putting my friends right upon that point.

O.—I am glad you have begun to ask questions, for that is the way to pick up useful knowledge. Allow me to ask, in what part of the world do you take up your abode during your absence from England?

N.—From August to April we seek the warm climes of the East, and are to be found as far as China and Japan.

O.—Do you visit all parts of England during your stay with us?

N.—We never venture on the north side of the river Trent, as the cold in that part of England would be too severe for us.

O.—I see you take the trouble to build a nest of dry grass and leaves. We lay our eggs in any hole or corner.

N.—Our eggs are of a very greenish brown colour; pray, what colour are yours?

O.—Ours are nearly white; and we generally lay five at a time, which I believe is one more than your number.

N.—You have been rightly informed.

O.—Hark! the cock crows. I must wish you good morning.

And away flew the owl to the old thatched barn.

THE PEACOCK AND
THE TURTLE-DOVE.

THE PEACOCK AND THE
TURTLE DOVE.

EVERY one admires the beauty of the peacock. The colour of his head, neck and breast, is deep blue, tinged with green and gold; his back is deep blue, with a shade of copper; and a little waving tuft adorns his head. But the train of the peacock is still more dazzling; at the end of each of the train feathers, which are of a dark colour, there is a spot like an eye, so that the peacock's train appears to be full of eyes. The middle feathers of it are sometimes more than fifty inches long.

One fine day, in July, on a newly mown lawn a peacock was walking at the stately pace at which these graceful birds usually move; he held back his head and neck a little to show himself off, and every now and then turned slowly around to catch the sunbeams on his splendid train, which was spread open like a fan. He was full grown, and must therefore have been three years old at least, for peacocks have not done growing before that age. He was also in full plumage, but that was no wonder as it was summer, the season in which peacocks wear their gayest attire. A little way off some peahens were feeding; they were much smaller than the cock, and looked quite plain compared with him. He was making his way towards them, when a dove suddenly settled on a low branch of a tree between him and the rays of the sun.

A dove is a kind of pigeon; a bird so well known that there is no occasion to tell the colour of all its feathers; but this bird had on each side of her neck a spot of black feathers tipped with white, which showed she was a turtle-dove, and her yellow eyes were set in a crimson ring.

The peacock was offended with the dove for casting a shadow on his train just when he was displaying it to gain the admiration of the peahens. In a very harsh tone of voice, (but that he could not help, for peacocks never have pleasant voices), he cried out, "How dare you throw your shadow on my train?"

The dove, instead of falling into a passion, and giving such answers as the game-cock gave to the turkey-cock, meekly replied, and the tone of her voice was soft and mild :

D.—I am sorry, Mr. Peacock, to have given you offence; I was not aware of what I was doing. I will be off this minute, for I stopped to admire, and not to eclipse your beauty.

P.—Pray do not go, Mrs. Dove, I spoke in too much haste; I am sure you did not mean to offend me.

D.—Although the day is so sultry, yet, at your request, I will stay a little while.

P.—It is indeed very hot; how is it that you are not in the deep shades of the forest, which is the usual resort of your tribe in such weather as this?

D.—To confess the truth, Mr. Peacock, I was tempted to leave my cool retreat to pay a visit to yonder pea-field.

And it is not unlikely that, besides peas, some wicken also might have been found in her crop, for doves are very fond of the weed which farmers call by that name.

P.—Pray, what further stay do you make in this country?

D.—We shall leave England in August, after having paid you a visit of nearly five months.

P.—You leave us then a month earlier, and come to us a month later than the wild pigeons?

D.—Yes; we cannot bear cold weather; when we leave England we seek the warm lands of the East.   I have been in India, whence the family of the Peacocks came to this country.   In that part of the world the plumage of most of the feathered tribes is much more brilliant than the plumage of those birds that are natives of colder climates.

P.—So I have heard. The Peacock family is a very old one; we are spoken of in the Bible as forming part of the treasures brought by the navy of King Solomon from Tharshish.

D.—And I dare say you remember the Dove is mentioned in the days of Noah.

P.—Of course, I do; and some of our family, doubtless, were with your ancestors in the Ark.

D.—Had you, my friend, lived at the time when the Romans had possession of England, it might have been your fate to have been eaten at one of their great feasts, for they thought a roasted peacock a great dainty.

P.—Those must have been sad times

for poor peacocks; but although I should not like to be eaten, I should not, after my death, object to have my train-feathers made into cloaks and fans for the ladies, as was the fashion in the reign of Queen Elizabeth.

D.—In the present day your feathers, I believe, are made no use of, except occasionally to ornament a chimney-glass. It is now time for me to seek my mate; I therefore wish you good day, Mr. Peacock.

P.—Good day, Mrs. Dove, I am happy to have made your acquaintance.

THE JACKDAW AND
THE RAVEN.

THE JACKDAW AND
THE RAVEN.

A BIRD which, from its size and colour,
might at a little distance be taken for a
crow by any one who did not know much
about birds, was perched on the very
top of the spire of a village church, and
and he was often to be seen there, for he
was a jackdaw, and it is well known that
these birds delight in perching on the
summit of the highest buildings, and
they are also fond of building their nests
in the towers of a cathedral, or of an
old abbey. The jackdaw is rather less
than the crow ; it has some white streaks
under the throat, and a few white dots

F

round the nostrils, and there are dots of an ash colour on the back-part of the head and neck. It has also a whitish ring round the black spot in the middle of the eye. The other parts of its body are quite black, but the upper part is a deeper black than the under part.

You may suppose, perhaps, that being perched so high above the ground, this jackdaw is thinking that all his bones will surely be broken if he should happen to fall; but no such thoughts enter his head. He is an active fellow, and like most active people, he is very cheerful. No one ever heard of a melancholy jackdaw. Seeing a raven on a high tree near him, and being tired of his own company, the jackdaw flew thither to have a chat with his cousin.

But before I tell you what they talked

about, let me give you some account of
the Raven. He is the largest of the Pie
family, to which both he and the jackdaw
are related. In length, the raven is
above two feet, and its breadth is four
feet. The upper part of its body is a
glossy black, with a tint of blue when
viewed in a particular light; the under
part is of a dusky hue. The raven's bill
is thick and strong; the nostrils, and
about half the bill, are covered with
hairs, which are nearly as strong as
bristles. The raven is bigger than the
crow or the rook.

The jackdaw being even smaller than
a crow, was, of course, a good deal smaller
than the raven. The jackdaw did not
perch very close to his cousin, but sat at
a little distance from him, and I will tell
you the reason. The jackdaw knew that

F 2

a very bad smell comes from the bodies of
ravens, which is owing to their feeding
on carrion ; they would be as sweet as
goldfinches and canaries if they eat
nothing but herbs and grain ; and it has
been proved that they can live on a veget-
able diet, so that there is no need for
them to eat horse-flesh and other nasty
things. But the truth is, ravens are
birds of a bad taste, and they are great
cowards into the bargain.   It is on this
account, probably, that they are ranked
as the lowest of the birds of prey.   It
was a matter of surprise to the jackdaw
to see the raven by himself, for he knew
that ravens are generally seen in pairs ;
jackdaws, on the contrary, often asso-
ciate in flocks.   Our jackdaw, in a lively
tone of voice, opened the conversation
with the raven, as follows :—

J.—I see, Ralph, you are in a moping mood, as usual. It is no wonder, with that grave face and solemn air, that people consider your presence as ominous, and in some countries even set a price on your head; and when you do open your mouth, in what a melancholy tone of voice you speak; but I ought to call it croaking rather than speaking.

R.—You ought to know, Jack, that although in some countries the ignorant doom us to destruction, yet in other parts of the world we are held sacred, and our motions are carefully watched by those who wish to know what is to happen to themselves or to others.

J.—That puts me in mind that I have heard there were people in ancient times foolish enough to eat the heart of a raven, thinking that after such a meal

they should be able to foretell future
events.

R.—Such things were done a great
many years ago, but folks are wiser now,
and would laugh at such prophets.

J.—It is well for you, cousin, that you
do not live in Greenland, for the natives
of that country eat the flesh of ravens,
and make clothing of their skin.

R.—I can assure you I have no wish
to emigrate to another country. In
England we are never eaten ; so far from
it, we are sometimes made companions
of by people who like our grave appearance
and sedate manners.

J.—I would undertake to fly to Greenland and back again before you got half
way thither; it is because you eat so
much meat, and take so little exercise,
that you are always dull and in low spirits.

R.—It does not become a bird who walks like a man, and does not hop along the ground as you do, to be continually in a bustle; and allow me to tell you, it would be better for your reputation, cousin, if you were to leave off stealing and hoarding up things which are of no use to you, but may be of service to their rightful owners.

J.—Cousin, you provoke me to tell you, it is very commonly reported that you watch for the shepherd's back being turned, and then pick out the eyes of the poor little lambs.

R.—I have no doubt that was told you by the rooks, who ought to be ashamed of themselves for telling family secrets. Although they are our cousins, they have neither affection nor friendship for us; and they have circulated a report about

you daws. They say that you destroy
the eggs of partridges.

J.—Come, Ralph, let us talk of something else, for it usually happens when
birds begin to find fault with one another,
such conversations end in pecking or
scratching.

R.—With all my heart: pray, Jack,
did you ever taste any kind of flesh?

J.—I have, now and then, taken a
mouthful of meat that came in my way,
just for a change in my usual diet; which,
you know, is fruit and grain.

R.—And I occasionally make a meal
on cherries, the only fruit I ever eat.

J.—As you ravens live on flesh, it is
fortunate for you that you have such
excellent noses, that you can smell carrion
half a mile off.

R.—I don't know whether you are

aware that we resemble you jackdaws in two remarkable particulars.

J.—Pray, let me hear them.

R.—Ravens as well as jackdaws are found in every part of the world ; and, like you, we do not wear exactly the same coloured suit in every country. In cold climates we lay aside our black, and put on a dress of a lighter colour. I have heard that in Norway jackdaws are some-times found quite white, and in Switzer-land some of them wear a white collar round their necks. I wish you would set that fashion in England; it must have a very pretty effect.

J.—There is one part of your conduct I have always admired. I allude to your constancy to one mate, and the care you take of her while she is sitting on. her eggs. You cock ravens, moreover, take your turn at sitting on them, which is very kind of you.

R.—By-the-by, that reminds me it is now January, the month in which we look out for a tree to build our nests in. Our favourite situation is a fork in the branch of an oak or an elm.

J.—We lay our eggs sometimes in rabbit burrows, and they afford us comfortable winter quarters, for we stay in England all the year round, as you do.

R.—We find the topmost branches of high trees to be very healthy situations, and the consequence is, we live much longer than other birds. Some of my ancestors lived to be a hundred. We lay five or six eggs of a pale bluish green colour, spotted with brown.

J.—We lay the same number of eggs, and they are smaller and paler than those of a crow. I dare say you know that, like you ravens, we are occasionally caught by boys, who teach us to speak a few words of a language which is quite foreign to us.

R.—I feel in much better spirits, Jack, than I did half an hour ago, and it is entirely owing to your cheerful conversation. If I were to have a chat with you every day, I should soon become as lively as a jackdaw. But you must excuse my leaving you now, cousin, for my nose tells me a pig has been killed in the village. I must hasten thither to get a bit for poor Spot the coach-dog, who is confined to his kennel by a broken leg. We are old friends, having been brought up together in the same stable-yard.

THE JAY AND THE WOODPECKER.

THE JAY AND THE WOODPECKER.

ONE fine summer's morning, a hen-jay, who was a great gossip, began as soon as it was light to look about for some one to chat with. Now, while she is waiting for some neighbour to pass by, let me tell you what sort of a bird a jay is.

The jay's head, neck, and back, are brownish red ; the wings light blue and white, with black bars ; and the tail is black. On its forehead there is a tuft of white feathers, dotted with black, and it can raise this tuft up, and let it down, as it pleases. The jay is of the pie family ; therefore the magpie, the rook, and the crow, are kin to it. It is one of the prettiest of our English birds.

The jay is a shy and cautious bird ; and, if disturbed, will fly from one tree to another, or from one copse to another, at the same time making a harsh chattering. And this bird also shows its caution in another way : it will just keep out of the range of gun-shot, as if it had learned by experience, or had been told by some one, that if it never allowed the shooter to come within a certain distance, it was safe.

When the jay had waited a short time, a woodpecker winged its way towards her.

J.—How d'ye do, this morning ?

W.—Quite well, I thank you.

J.—Is there any news stirring ?

W.—Gossip Jay, I cannot stay to

chat, for my little folks at home are waiting for their breakfast.

This thoughtful woodpecker was of a green colour, and about the size of the jay; for the green ones are the largest of the woodpecker tribe. The woodpecker's bill is formed like a wedge, and its neck being short and thick, it can strike its bill with so much force against the trunk of a tree that the sound may be heard a good way off. The noise disturbs the insects under the bark, they come out to see what is the matter, and the next moment they are in the woodpecker's crop. This habit of pecking trees gives the bird its name. And, besides this, the woodpecker's tongue is so long and taper, that it can thrust it into many of the small holes which are found in the bark of trees, and bring out on the tip

G

of it the insects which live in those small cracks ; and there is no food the woodpecker likes so well as insects.

The merry laughing cry of the woodpecker may often be heard in the neighbourhood of woods, or issuing from some large tree, to the trunk of which the bird may be seen clinging, with its head thrown a little back, the weight of the body resting almost entirely upon the tail, the feathers of which are hard and wiry, more like bristles than feathers ; and in old woodpeckers they are generally much worn at the edges, by being constantly rubbed against the rough bark of the trees. The woodpecker is a shy and solitary bird, but yet it is sometimes tamed.

Soon afterwards, a rook happened to fly by.

J.—Good morning, cousin Rook.

R.—The same to you, Jenny Jay.

J.—Pray, stop a minute.

R.—I really have no time to spare; for I must collect some sticks to mend my nest, which was nearly blown down by last night's high wind.

And the rook kept on his way.

J. (*solus*)—I think Master Rook might have chatted with me half an hour. My own nest is out of repair, but there is plenty of time between this and winter to mend it.

Now, the jay would not have had much trouble had she set about mending her nest, for it was made only of twigs. Jays lay five or six greyish eggs, spotted with brown.

The jay flew from tree to tree, for it is a restless bird, turning her eyes on all sides, to find some one as idle as herself. She now espied, in a thicket hard by, a nightingale, to whom, without loss of time, she addressed herself.

J.—I hope I see you well, Mr. Nightingale. What a fine day we have after last night's storm.

The nightingale nodded assent, and kept on picking the berries from the bush upon which he was perched. The jay, not at all disheartened by this kind of answer, continued—

J.—If you will sing to me one of your charming songs, for I know you sing by day as well as by night, I will, in return, tell you what news there is among the finches, pies, and sparrows, of these parts.

The jay said to herself, though Mr. Nightingale will not talk, he will perhaps

sing. Instead, however, of complying
with her request, he answered—

N.—If I sing now, Mrs. Jay, I shall not
be in good voice for the concert, at which
I have promised to sing, this evening.

Before the jay could reply, the songster
had taken wing.

J. (*solus*)—Well, I should only have
had a song from him if he had staid;
and for my part, I much prefer talking
to singing. What were our tongues
made for, but to talk with?

The jay's voice was very harsh, and it
is no wonder that it was so, when we
consider that she did nothing but chatter
from morning till night.

In the course of the day, a great
many birds of all colours and sizes
passed within the reach of her voice;
and with each of them she would have

liked very much to have had a gossip, but they all had something better to do than to talk about the weather and their neighbours' affairs. Some of them, indeed, gave her very short answers. Thus this gossiping jay spent the fine long days of summer. Winter at length set in; and this foolish bird having taken but little food when it was easy to be got—for, as we have seen, she used her mouth more for talking than for eating—she had grown so weak, that she was not able either to mend her nest, or to migrate to another country, as a great many jays do every year. Had a hawk attacked her she must have fallen a prey to him. She had not strength even to get her daily food; and had it not been for the charity of her neighbours she must have died of cold and hunger.

THE STARLING AND
THE LINNET.

THE STARLING AND
THE LINNET.

A BIRD, which at a little distance might be taken for a small blackbird, was hopping about a meadow one summer's day at noon, but it might have been seen there in the winter, for it was a starling, and the starlings do not leave us in cold weather, but are our guests all the year round. It is seldom you see one of these birds by itself, for the starlings live in flocks, besides they are often to be seen in the company of crows, rooks, and jackdaws, with all of whom they are on very friendly terms. The plumage of the starling is of a dark colour, spangled with green, blue, purple, and copper; and at the end of each feather there is a pale yellow spot. The starling chatters most in the evening and morning, but when it pleases it can talk just as well at other times, as you will find by what passed on the present occasion between this starling and a linnet who happened to be near him. You should, however, first notice the plumage of this pretty little bird. The upper part of the linnet's head, neck, and back, is reddish brown, and the under parts reddish white. Its sides are streaked with brown. The breast of this linnet was crimson, which showed he was a cock, for hen-linnets have streaks of brown on their breast.

The Starling and the Linnet.   P. 107.

S.—Pray, have you dined, neighbour Linnet?

L.—I have not yet had my dinner, but I see it is dinner time, for the sun is full south, and noon is our dinner hour. I have been at a morning concert in yonder thicket, and was not aware it was so late.

S.—I have been paying a morning visit to Mr. and Mrs. Crow, who were so kind as to ask me to take pot-luck with them to-day. I dare say they have something nice for dinner, and as they are very old friends of mine, I will, if you please, take you with me, and you may depend on receiving a hearty welcome.

The linnet being hungry, accepted the starling's offer, and on their way to Crow Hall they chatted as follows :—

S.—I believe, Mr. Linnet, your name comes from your favourite food, linseed.

L.—Some, people suppose so, but others doubt it. Talking of names, how strange it is that in some parts of England starlings are called stares. I do not remember to have heard what your usual food is.

S.—We live chiefly on worms, snails, and grubs.

L.—I have been told there is one kind of fruit you are very fond of.

S.—We like cherries much better than any other fruit. I hope you will excuse what I am going to say. I have often thought it hard that you gentlemen linnets should be not only more finely drest than your ladies, but also that you should be able to sing, while they can only twitter.

L.—Alas! we often pay a penalty for that superiority. Nets are spread upon the ground to catch us, and when we are caught we are shut up in a cage for the rest of our days, although our song is not nearly so good then as it is in the open air; and, besides this, the crimson hue on our breasts fades away.

S.—Even starlings are sometimes put into cages, although we have no song at all. Silly boys try to teach us to speak, but after all their pains we make sad work of the Queen's English.

L.—One evening last winter I saw hundreds of starlings flying in the air in a close body, and wheeling about like soldiers, when all at once they settled on the ground out of my sight, but I could still hear them chatter, and not a little noise did they make with their tongues.

S.—When we meet in large numbers we look out for a bed of reeds, and roost among them for the night. But I am afraid it will be a long time before you will see such a collection of starlings, for our numbers are lessening every year in England.

L.—I wonder you are not afraid of roosting in such damp places as reed-beds usually are.

S.—We are in the habit of bathing, and are therefore not very particular as to the dryness of our roosting-places.

L.—Do you take much pains about your nests?

S.—Very little indeed, any hollow place serves us to lay our four or five ash-coloured eggs in; but we take care not to let passers-by see where we lay them.

L.—We build our nests of grass and moss, in low bushes, and we lay the same number of eggs that you do, but ours are pale blue, spotted with brown at the larger end.

The starling now cried out, " here we are at my friends' house." Mr. and Mrs. Crow made the linnet welcome, as the friend of Mr. Starling. They had also asked their cousins, the Rooks and the Daws, to dine with them. The dinner, which was a large piece of horse-flesh, was served up under an old oak. "You see your dinner," cried the hostess, "pray begin." The words were no sooner out of her mouth than all the guests, except the linnet, began with all their might and main to pull the meat in pieces. The linnet, with surprise, saw them scramble for the choicest pieces, and it ended, as scrambles generally do, in downright quarrelling. They scratched and pecked one another, and in other ways showed how angry they were. When there was no more left to eat, the guests took their leave. The starling and linnet went away together, and on their way home the following conversation passed between them :

S.—How did you like my friends' dinner?

L.—Why, to tell you the truth, Mr. and Mrs. Crow's cousins were at first so hungry, and afterwards grew so quarrelsome, that I did not get a single mouthful.

S.—I am very sorry to hear you say so, for I never tasted a bit of better horse-flesh in my life.

L.—Its flavour might be very good,

but I hope never to be present at such a dinner again. I prefer coarse fare, eaten in peace, to the greatest dainties, where strife and ill-will prevail.

THE PHEASANT AND
THE WREN.

THE PHEASANT AND
THE WREN.

ONE bright, calm day, towards the end
of autumn, there was perched on a tree
by the side of a beech-wood a bird with
a head and neck of deep purple; the
breast chesnut, and the back speckled
with black, white, and orange. Its tail
was very long and handsome; the two
middle feathers being much longer than
the other tail feathers, which were reddish
brown, crossed with small black bars.
Around its eyes there was a scarlet ring.
This bird was rather less than a common
cock. Every one who knows any thing

about birds would at once say it was a cock pheasant.

The pheasant spied, just below him, a very small bird, of a dull brown colour, with a little gray on the under side of its body. At a short distance it might have been taken for a tiny wood-cock. It was a wren, that brisk little bird which hops on our garden gates in winter, when its loud and clear note is often heard at the close of the day. And it is also known by its whirling kind of flight.

The pheasant looked down with contempt on the poor little wren. He tossed his head, as much as to say, I do not think you worthy of my notice. The wren paid no regard to the airs which this proud bird gave himself, but hopped from twig to twig, just as she would have done had no such grandee been near her. This hurt the pride of the pheasant, who thought, because he was so much bigger and finer than the wren, that she ought to have treated him with more respect. He could no longer refrain from speaking.

P.—How is it that you do not show your sense of the honor I do you by being seen on the same tree with such an insignificant bird as you are? Has my great beauty struck you dumb with wonder?

W.—I did not think, sir, it became me to speak to one whose rank is so much above my own.

P.—I am glad you know the distance there is between the king of the British woods and a paltry wren.

W.—It is very well known, sir, that I am generally to be seen in company with the robins, who are only a little bigger than wrens.

P.—I dislike those robin red-breasts.

W.—You surprise me; I find them very good neighbours.

P.—I own that I am not of a social temper, but it is not on that account I dislike them; it is the colour of their waistcoats that offends me.

Pheasants as well as cock-turkies dislike every thing of a red colour.

W.—That is strange, if, as I have been told, you were brought to this country from the eastern part of the world, where there are many red and scarlet birds.

P.—What you have been told is true; but perhaps you may not know that we have not yet visited the new world.

It is a fact that there are no pheasants in America.

W.—As I have never happened to see pheasants feeding, pray may I ask what kind of food do you prefer? We wrens live chiefly on dead insects, when we can get them.

P.—We are fond of all sorts of grain, and we also eat ants with a relish, but there is nothing we like better than white peas.

W.—Your nests, I dare say, differ from ours, which are sometimes taken for lumps of moss, they are lined with feathers, and we lay nine or ten eggs, of a dirty white colour, with reddish spots.

P.—We take little pains with our nests, but we lay twice as many eggs as you do, and they are of a brownish colour. We also differ from you in our mode of sleeping; we do not rest our heads upon our breasts, as most birds do, but we tuck them under our wings, and so we sleep very soundly.

W.—As you sleep so well, are you not in danger of falling a prey to some of your enemies?

P.—Alas! when we are at roost we are often caught by the poachers, who certainly have no right to take our lives, as we do not live at their cost, but at the expense of those whose corn we eat.

W.—I know not which to admire most, the beauty of your plumage or the length of your tail.

P.—Talking of tails, I observe you always carry yours cocked up. I suppose you carry it so because it is not more than an inch long, and you wish to make the most of it.

W.—Nature has kindly made up for the shortness of my tail by giving me a cheerful voice.

It would not have been good manners in the wren to have told the pheasant that his note was a harsh scream, although had she said so she would have said no more than was true.

P.—But your suit is as plain as a Quaker's, while I am drest as fine as a courtier. Do you not wish you were as large and as fine as I am?

W.—I am quite contented to be plain Jenny Wren, for we little birds are not so liable to accidents as you fine large birds are.

P.—What a mean spirit you have; but to be sure, humility is as proper in a wren, as pride is in a bird of my rank and beauty.

The report of a gun was at that moment heard, and the proud cock-pheasant dropped from the tree dead.

CAMDEN PRESS, LONDON.

A NEW AND ATTRACTIVE SERIES OF
JUVENILE BOOKS,
EACH VOLUME ILLUSTRATED WITH COLOURED ENGRAVINGS.

———————◆———————

THE ROSE-BUD STORIES.

A Series of Illustrated Volumes (sixteen varieties) uniform in size and style, price One Shilling and Sixpence each. Every volume contains one or more Tales complete, is strongly bound in cloth boards, with four coloured engravings on wood, designed and engraved by Dalziel Brothers, and 124 pages of clear, bold letter-press, printed upon stout paper.

The Tales are written by various Authors, most of them expressly for the Series, and for cheapness, attractiveness, and sterling interest, they present, perhaps, one of the most pleasing and useful collections of Stories in modern Juvenile Literature.

———————————

As the fresh Rose-bud needs the silvery shower,
The golden sunshine, and the pearly dew,
The joyous day with all its changes new,
Ere it can bloom into the perfect flower;
So with the human rose-bud; from sweet airs
Of heaven will fragrant purity be caught,
And influences benign of tender thought
Inform the soul, like angels, unawares.

MARY HOWITT.

[*For Contents, see following pages.*

[*For Contents, see following pages.*

LONDON: JAMES HOGG & SONS.

I.

ALLY AND HER SCHOOL-FELLOW.
A TALE FOR THE YOUNG.

By MISS M. BETHAM-EDWARDS,
Author of " Holidays among the Mountains," " Charlie and Ernest," &c.

II.

LOYAL CHARLIE BENTHAM.
By MRS. WEBB,
AUTHOR OF " NAOMI," ETC.

AND

THE CHILDREN'S ISLAND;
A TRUE STORY.
EDITED BY L. NUGENT.

III.

SIMPLE STORIES FOR CHILDREN.
By MARY E. MILLS.

IV.

A CHILD'S
FIRST BOOK ABOUT BIRDS.
BY A COUNTRY CLERGYMAN.

V.

PRINCE ARTHUR; OR, THE FOUR TRIALS.
By CATHERINE MARY STIRLING.

AND

TALES BY THE FLOWERS.
By CAROLINE B. TEMPLER.

LONDON: JAMES HOGG & SONS.

LONDON: JAMES HOGG & SONS.

LONDON: JAMES HOGG & SONS.